BIOSECURITY GUIDE FOR
LIVE POULTRY MARKETS

FOOD AND AGRICULTURE ORGANIZATION OF THE UNITED NATIONS
Rome, 2015

Recommended citation
FAO. 2015 *Biosecurity guide for live poultry markets.*
FAO Animal Production and Health Guidelines No. 17. Rome, Italy.

The designations employed and the presentation of material in this information product do not imply the expression of any opinion whatsoever on the part of the Food and Agriculture Organization of the United Nations (FAO) concerning the legal or development status of any country, territory, city or area or of its authorities, or concerning the delimitation of its frontiers or boundaries. The mention of specific companies or products of manufacturers, whether or not these have been patented, does not imply that these have been endorsed or recommended by FAO in preference to others of a similar nature that are not mentioned.

The views expressed in this information product are those of the author(s) and do not necessarily reflect the views or policies of FAO.

ISBN 978-92-5-108910-1

Contents

Acknowledgements

This guide builds upon the body of work conducted by the Food and Agriculture Organization of the United Nations (FAO) and other development partners, the World Bank and the World Health Organization (WHO), and by country collaborators in improving live poultry market management in terms of hygiene and consumer safety. Options for change and ensuing recommendations on live bird marketing practices were made initially once the importance of live poultry markets was recognised as a key intervention effort in halting the persistence and potential transmission of zoonotic avian influenza viruses in Hong Kong in 1997, but adaptation to other environments and countries were required with the further spread of the H5N1 influenza strain to more than 10 countries in early 2004 and to poultry producing systems in over 40 countries by 2006. Much of the information in this guide builds on experiences of Dr Les Sims working in a number of countries that have made changes to live poultry markets over the past 18 years, especially Cambodia, People's Republic of China, and Viet Nam. A debt of gratitude is owed to the many people who have assisted the production team in building an understanding of how markets operate; behaviours of sellers and buyers; on issues encountered in making changes to markets and marketing practices and also in providing access to markets; and a myriad of intermediaries. Special thanks go to Lai Thi Kim Lan, John Weaver, Sorn San, John Edwards and Guo Fusheng. The Annex to this document on market cleaning and decontamination was a results from material provided to FAO by Andrew Almond. Very useful discussions were held with Sue Trock on market decontamination and use of disinfectants, based on experiences in markets in the United States of America. A number of photographs used in this guide were provided by FAO staff with special thanks to Astrid Tripodi and Charles Bebay, who also provided valuable information from African markets. Special thanks to Cecilia Murguia for coordinating the editorial and design work, Enrico Masci for design, Shannon Russell for style editing, the FAO peer-reviewers, Paolo Pagani, Charles Bebay and Fallou Guèye, for their feedback on the content of the guidelines, as well as Sophie Von Dobschuetz and Astrid Tripodi for managing its development.

This guide was made possible through financial support provided by the United States Agency for International Development (USAID). The opinions expressed herein are those of the author and do not necessarily reflect the views of USAID.

Gratitude is re-extended to Les Sims, whose insight and work during and since his interventions in 1997 are collated in this guide.

Acronyms

LBM   Live Bird Market

PPE   Personal Protective Equipment

WHO   World Health Organization

Introduction

This guide provides options for improving biosecurity in live poultry markets, focusing on those areas that will have the greatest impact. The guide examines individual scenarios and works on how to improve market biosecurity so as to reduce the risk of disease spreading from bird to bird, or from birds to humans.

Live poultry markets have been implicated in the zoonotic transmission of avian influenza viruses from live poultry to people (traders and customers). This transmission can occur via direct or indirect contact although, in many cases, the exact route is not known.

Avian to human transmission of avian influenza viruses in markets was first recognized with influenza A(H5N1) viruses from 1997 onwards. It has been complicated by the emergence of a range of other zoonotic influenza A virus subtypes, including a virus of the H7N9 subtype. As of April 2014, influenza A(H7N9) has caused more than 470 cases of disease in humans since it first appeared in China in March 2013. The H7N9 subtype virus has placed neighbouring countries and trading partners at risk.

Measures need to be taken in live poultry markets to reduce the likelihood of infection of poultry and people, with these viruses.

It should be viewed as a failure of control and preventive systems if any person working in or visiting markets is infected with influenza viruses derived from poultry, and develops severe disease.

Therefore, regardless of the market type, the over-riding objective is for poultry sold in the market, and the market environment, to pose minimal threat to the health of market traders and customers. This objective has been achieved in places such as Hong Kong SAR and Shanghai where a range of measures (described below), some of which are costly to implement and enforce, have been introduced that have prevented loss of life from zoonotic avian influenza.

In reality, the best that can be achieved in many other places is to reduce this risk by implementing practical low-cost measures that reduce viral loads. This is especially the case in resource-poor cities and countries with a long history of trade in live poultry and established markets and practices. In many existing markets or places where poultry transactions occur the facilities are such that they are difficult to clean and disinfect.

The World Health Organization (WHO) has pointed out that, in many cases, "markets are settings with a very low level of resource generation and therefore have little financial resources for maintenance and improvements. It is therefore extremely important to identify issues that pose a real public health risk and which need immediate attention rather than issues that would be nice to address but which do not pose a clear documented public health risk."[1]

The limited availability of resources to remedy problems needs to be kept in mind when developing interventions. There are also many different types of markets so it is

---

[1] See http://www.who.int/foodsafety/fs_management/live_markets/en/.

not possible to develop a single set of standards covering all live poultry markets for all countries. Markets range from local markets at a commune or district level that open intermittently and may only sell 10 to 50 live birds a day, to large wholesale markets with a daily throughput of tens of thousands of poultry, to large retail markets with multiple stalls where hundreds of birds of different types are sold, slaughtered and butchered on a daily basis with poultry kept overnight and no poultry-free market 'rest days.'

The guide provides information on reducing the risk of introducing zoonotic avian influenza viruses to markets, as well as their propagation and onward transmission to other markets or other places, including restaurants and homes.

There have been other guidelines written on market hygiene and these should be referred to for additional information (see Selected references for some examples).

Descriptions of different market situations and ways to resolve issues

Market managers can use this section to find situations that are directly relevant to the markets they manage.

❝ I AM RESPONSIBLE FOR MARKETS IN A MAJOR CITY AND I NEED TO MAKE A DECISION ON WHETHER TO KEEP LIVE POULTRY MARKETS OPEN OR SWITCH TO CENTRALIZED SLAUGHTER ❞

Those in charge of poultry marketing in all major cities should make a decision on whether live poultry sales are required, whether a shift towards centralized slaughtering will occur and the extent of the reduction in live poultry sales that is required. In some places, a decision will be made to have most of the product sold as chilled, fresh chicken via retail outlets (including wet markets), therefore reducing the volume of trade in live poultry, and the number and size of live poultry markets. Such moves must be feasible, acceptable to the public and the trade, and should weigh up public health benefits and consumer preferences. Appropriate policies and resources for implementation, and strict enforcement against illegal trading must support these changes.[1]

This is the situation in Hong Kong SAR in 2015 where chilled and frozen poultry are supplied from mainland China and elsewhere, and live poultry sales in retail markets represent only a small proportion of the total daily trade in poultry. It is being proposed and trialled in other parts of southern China such as Guangdong province, and has been adopted in Shanghai. Other places have imposed a total ban on the sale of live poultry in urban districts such as Ho Chi Minh City in Viet Nam and Ningbo in Zhejiang province of China to reduce the risk of zoonotic transmission of avian influenza viruses. It is not always easy to achieve this goal because of consumer preference for freshly slaughtered poultry.

In relatively rich cities with well-developed food supply networks and refrigeration, some of the reasons for relying on live poultry sales are less compelling than they were 20 to 30 years ago. In the past, customers had no access to refrigeration and wanted to see a healthy bird because it was a sign that it would not pose a threat to their health. However, with zoonotic avian influenza viruses – especially, but not only, influenza A(H7N9) virus – even if a bird appears healthy, the possibility that it is shedding virus, thereby putting both traders and customers at risk, can not be ruled out.

---

[1] Permanent closure of all live bird markets (LBMs) is not necessary if the markets can be managed appropriately with strict control on the sources of birds and implementation of measures such as those described in this document. Permanent closure should be considered only when alternative, hygienic venues for trade and slaughter are available, such as well-managed poultry slaughterhouses in which measures are taken to protect slaughterhouse workers from infection with zoonotic influenza viruses. Before deciding to close any LBM permanently, the competent authorities should consider the impact on livelihoods and, where possible, should provide support in finding alternative employment for people no longer able to continue their trade.

Other reasons for relying on live bird sales are that 'on-the-spot' butchering means product substitution is less likely (with, for example, a bird that had already died). Some consumers also prefer the taste and the texture of freshly slaughtered poultry compared to chilled or frozen product.

Experiences in Hong Kong SAR have shown that consumers and restaurants can adapt to availability of only chilled or frozen product, as has been the case with the supply of ducks and geese since 1998 when these birds were no longer sold live in retail markets. Increasingly, the restaurant trade in Hong Kong SAR has switched to the use of chilled or frozen chicken and chicken products imported from mainland China and elsewhere (because the supply is reliable and it is cheaper than live poultry). Nevertheless, there is still a demand for freshly slaughtered chicken and other types of poultry in Hong Kong SAR and this is delivered via a very tightly controlled supply chain, starting from farms through to transportation and to both wholesale and retail markets. Buying habits of most consumers in Ho Chi Minh City in Viet Nam have also changed given that live poultry sales in urban markets were banned following a shift to centralized slaughter around 2005.

The retention of some well-regulated live poultry markets reduces the risk that the live poultry trade will be driven underground to illegal, unregulated sites, or to peri-urban markets, especially if there continues to be demand for the product.

All countries that choose to sell live poultry in and around 'wealthy' urban areas should be aiming for the standards described below to minimize the risk to consumers.

In poorer cities and many rural areas, it is not possible to attain these standards in most markets until changes occur to the systems of production and trade. Major changes to facilities in markets may also be required to meet these 'platinum' standards.

Any decision on retention of live poultry markets must be made in conjunction with public health officials.

❝ I HAVE DECIDED TO MAINTAIN A SMALL NUMBER OF WELL-MANAGED WHOLESALE AND RETAIL LIVE BIRD MARKETS (LBMS) THAT POSE MINIMAL RISK TO THE PUBLIC ❞

The following standards should be met for well-managed markets that will pose minimal risk to human health.

High-level (platinum) standards for wholesale markets selling live poultry

Market location, design and facilities:
- Markets are located at least 200 metres from the nearest residential area in zones with no plans for residential development.
- There are separate markets for terrestrial and aquatic poultry; (ideally, on separate sites).
- All surfaces are non-porous and/or can be easily cleaned and disinfected, such as tiled floors and walls, and metal or plastic cages.
- Poultry are kept off the ground, either in cages or on raised non-porous matting, to reduce faecal contamination of poultry and the market;
- There are adequate supplies of potable water and equipment for cleaning.

- Appropriate facilities allow anyone entering or leaving the market to clean footwear (i.e. scrubbing brush, detergent, water and disinfectant).
- Handwashing facilities are in place with appropriate signage and training on use.
- Cages are designed and positioned so as to facilitate regular cleaning of areas under the cages, including removal of faecal matter.
- Truck- and cage-washing facilities are available on-site or close to it, and all vehicles and cages leaving the market are cleaned before departure (or sent directly to an off-site cleaning area if suitable facilities are not available in the market).
- Well-designed drainage (covered drains) must be in place with holding tanks for treatment of liquid waste before discharge and with removable traps for collection of solid waste that are emptied regularly to avoid blockage of drains.
- Systems for handling solid waste are in place that do not pose a risk to public health or poultry health (see Annex 1).
- Appropriate systems are in place to treat liquid waste.
- Systems for handling dead poultry are in place that do not pose a risk to public health or poultry health.
- A holding area is available for any suspect consignments (e.g. poultry without appropriate certification or showing signs of disease or having suspicious test results).
- Separate entry and exit points should be provided for vehicles bringing poultry into and out of the market.

Traceability:
- There is control over entry of all poultry with all batches of poultry accompanied by reliable and relevant health certification, preferably backed by appropriate laboratory testing. If vaccination for any strain of zoonotic avian influenza is mandatory, the relevant vaccination history must be provided.
- All farms supplying poultry to the market should meet agreed biosecurity standards.
- Permission for vehicles to enter the market is obtained prior to arrival so that market management expects all consignments before they arrive (e.g. SMS or phone booking system for booking in consignments, linked to licence plates of delivery vehicles).
- All birds in a single consignment are from one source and come directly from farm to market with no mixing of poultry types in consignments.
- There should be no official parallel channels to retail markets in the area covered by the wholesale market(s), other than retail market stalls that source poultry directly from specified, certified farms allowing for traceability (see retail markets below).
- All poultry leaving the wholesale market can be traced back to the farm of origin (e.g. docket system from wholesalers to retailers linked to certificates accompanying poultry from the farm of origin to the wholesale market).
- Poultry leaving the market can only go to retail markets, to slaughterhouses or to restaurants with appropriate holding and slaughter facilities.
- Records must be kept by market authorities and by individual traders of all incoming and outgoing poultry.

Market management:
- No mixing of poultry species in market stalls is permitted.

- No wild birds should be sold or kept in the market.
- No mammals should be sold in the market.
- Poultry are transported to and from the market in trucks and cages that after each use are cleaned, disinfected, inspected and certified after each use.
- Markets may be privately or publicly operated.
- All markets must have strong enforcement of biosecurity standards by veterinary and public health authorities.
- Traders must support and be aware of the importance of the measures being implemented.
- The market should be regarded as a sorting/selling point for rapid onward movement of poultry to slaughter or to retail markets, not as a holding area for poultry.
- Poultry are not to be kept on-site for more than 24 hours.
- Under exceptional circumstances, due to fluctuations in demand, a small number of poultry kept in appropriate cages away from any incoming poultry arriving the following day, may be allowed. In such cases, any bird kept for 24 hours must be sold or removed on the second day as a priority.
- Regular (at least once per month) poultry-free rest days are implemented with the market closed for a minimum of 18 to 24 hours, during which additional cleaning and disinfection are conducted (see below for details on rest days).
- Appropriate rodent and insect control programmes that pose no hazard to human or poultry health must be in place.
- No slaughter of poultry should occur within the market, (but the market may be connected to a slaughter plant).
- No vehicle arriving with poultry can leave with empty cages unless the vehicle and cages have been cleaned and disinfected or are going directly to an off-site cleaning area.
- No vehicle can reload with poultry after unloading until such time as the vehicle has been cleaned, and clean and decontaminated cages are available.
- Disposal bins for dead poultry are available with regular dead bird testing.
- A reporting system is in place for multiple dead poultry.
- These management procedures are covered by a set of standard operating procedures that are well understood and followed by traders and market management personnel.
- Market managers understand value chains related to their market.
- Restrictions may be placed on entry of poultry from certain areas if zoonotic influenza viruses are known to be circulating in these areas.

Volume of trade:
- The number of poultry traded through the market should be restricted to a level that allows full tracing in and out of the market and does not result in keeping of poultry for longer than 24 hours other than in exceptional circumstances (see procedures for handling excess poultry).
- The larger the market, the greater the consequences if a zoonotic agent is detected resulting in temporary market closure. Therefore, it is preferable to have multiple smaller and structured markets than one very large market trading more than 100 000 heads of poultry per day.

- There must be sufficient manpower to inspect, clear and record all consignments at peak entry and exit times.

High-level (platinum) standards for retail markets selling live poultry
- No overnight keeping of poultry is permitted.
- Thorough cleaning is done every night once the market is closed.
- No live poultry can leave the market.
- There is separation of customers from live poultry via a barrier system (e.g. poultry are behind glass or other forms of effective separation).
- There is separation of poultry display areas from slaughter areas preferably with one-directional flow from live poultry to dressed carcasses.
- There is appropriate ventilation of the area housing poultry and slaughtering areas that does not result in exposure of market customers to air from these areas.
- The poultry section is separated from other stalls in the market with no unnecessary through traffic.
- Ideally, the retail market should have a separate unloading area for poultry that is not used by other vehicles delivering goods to the market. Regardless, the area where consignments are unloaded must be cleaned after each batch of poultry is delivered. Also, that area should not be accessible by the general public.
- Poultry come directly from wholesale markets or from certified farms with a certificate of origin.
- If possible, batches from one source are held and processed together (but this is not consistent with work practices for most retail markets).
- Appropriate rodent and insect control programmes are in place.
- Multiple sites are available for customers and traders to wash their hands with soap and water, accompanied by appropriate signage.
- Appropriate drainage system, liquid waste and solid waste handling systems are in place.
- No mammals or wild birds are to be sold or kept in this section of the market.
- Any sick birds should be reported to the frontline livestock/veterinary officer and/or the market authority.
- All dead birds should be reported to the market authority and frontline livestock/ veterinary officer.

The measures above represent the highest standard for markets in countries or places that have the resources to implement these measures.

This type of marketing system does not cater readily for smallholders unless they can develop and sell specific, certified products, as has occurred on a small scale in Viet Nam. It is not suitable for traders and aggregators who purchase poultry from multiple sites.

" I NEED TO RETAIN LIVE POULTRY SALES BUT HAVE LIMITED RESOURCES AND CANNOT IMPLEMENT ALL OF THE MEASURES LISTED IN THE PREVIOUS SECTIONS "
It is not possible to provide rigid rules for all market types largely because resource limitations will prevent some measures from being applied, and the facilities in the markets are such that

even basic measures such as cleaning are extremely difficult to implement. Nevertheless, it is possible to make changes to many areas in the market, even in places with limited resources.

These measures can only be implemented with the support of local authorities, market management and market traders. Gaining support may require well-developed communication strategies that demonstrate to traders how the benefits of changes outweigh the costs. In some countries, market traders and local officials saw little reason to change existing practices even after the emergence of influenza A(H5N1), which was a potential threat to their health. In part, this resistence resulted because there were few human cases despite the intermittent presence of H5N1 virus in markets. Unless local decision-makers, traders and managers are committed to and understand the reason for the measures, they will not be implemented properly. Market managers and local authorities also need appropriate powers to enforce required practices.

For example, if a compulsory rest day is introduced, but cooperation is poor and enforcement is weak, poultry remaining in a market just prior to the rest day will not be slaughtered. Instead, birds will be moved to another site and then returned after the rest day, (or they will be left in the market). This kind of evasive behaviour is to be avoided. Cooperation is more likely if negative economic effects result from the presence of a particular strain of virus. A good example is influenza A(H7N9) virus in China which has resulted in prolonged closures of markets in some cities. If these closures can be avoided by having regular rest days, traders would see the advantage of cooperating. It is, therefore, important to ensure that traders understand the reasons for, and are supportive of, any measures introduced prior to their implementation.

Care needs to be taken that costs to traders do not increase. Otherwise, they might move to other sites or not cooperate. Nevertheless, places with poor facilities that do not allow rudimentary cleaning because of unsealed surfaces should be improved or, eventually, closed. However, any market closures will be in vain unless authorities have appropriate powers and resources to prevent the development of illegal parallel markets.

All of the situations listed below by market traders are relevant to biosecurity in many markets. The answers are aimed at reducing the likelihood of zoonotic avian influenza being introduced to, multiplying and persisting in markets with onward transmission to other poultry and humans.

❝ POULTRY IN MY MARKET ARE SOLD WITH OTHER GOODS IN A MARKET WITH NO SEPARATION ❞

Live poultry sections in markets should be moved to areas away from other stalls to minimize unnecessary through traffic, as in the picture below.

Note that cages and poultry in this market are placed over a tiled surface with appropriate drains that can be easily cleaned. Customer walkways are not contaminated and customers are kept a short distance from poultry. Customers do not have to walk through this section to access other areas of the market.

Insufficient separation of poultry from other raw food in a market.

The poultry market in this picture has been separated from market stalls selling other produce.

❝ VENTILATION IN MY MARKET IS POOR ❞

The following three examples show markets with poor ventilation.

Note also the wooden slats that are hard to clean and the uneven surface that is difficult to decontaminate effectively.

Air flow in the rear of the stall (where slaughtering is performed) is poor.

The slaughter area in this market is surrounded by live poultry, and ventilation is poor.

This market was fully enclosed and has poor ventilation especially in the middle of the market with the only available air flow from the entrances.

Areas where poultry are sold and processed should be well-ventilated as in the following picture.

LES SIMS

A market with good ventilation.

The market pictured is well-ventilated, easy to clean and separated from other activities in the market. Traders and poultry are protected from the weather. Note: This market could be improved by the provision of elevated cages or slatted elevated floors so that poultry are not kept on the concrete or on hay. Markets should have regular rest days (easy to implement for markets that open once every few days or where birds are not kept overnight).

Areas where poultry are slaughtered should be separated from the public. Ideally, markets should be designed in a way that segregates areas where poultry are delivered from areas where poultry are kept and displayed for sale. Areas where poultry are slaughtered should not be accessible to the public and should be separate from places where dressed poultry are offered for sale. In many markets, these requirements necessitate the redesign and reconstruction of stalls (see platinum standards for retail markets).

“ POULTRY ARE SOLD ON THE GROUND, WHICH GETS VERY MUDDY IN THE WET SEASON ”

Where possible, surfaces on which poultry are housed should be impermeable. Areas such as those in the photos cannot be properly cleaned and disinfected after trading. Measures need to be taken to discourage trade of this kind by providing appropriate places for marketing, and illegal sales need to be phased out once appropriate low-cost improvements are made. Resistance to change is likely to be high if practices are long standing, and parallel market channels will develop unless there is appropriate regulation and enforcement. Ultimately, a shift requires the support of traders who will need to be engaged in the process, who understand the reasons for any changes in practices and can afford any additional management fees or other costs.

FAO/NGUYEN THI THANTHUY

AMR ABDALLAH

Poultry sold on the ground.

Until such markets can be upgraded, it may be possible to remove some solid waste from the surface of the soil/mud and to use either vinegar or oxidizing agents (such as Virkon-S powder) on the areas where poultry cages have been placed. Systems are needed for washing of poultry cages, but in temporary markets such as these, the cages would usually go back to households.

❝ FACILITIES IN MY MARKET ARE IN POOR CONDITION ❞

Facilities in poor condition may have surfaces that are cracked and broken, drains that are not covered, walls and floors that are not sealed, or cages that are in a poor state of repair.

Cages in poor repair and constructed of material that is difficult to clean.

Slaughter area that is difficult to clean due to the materials used in its construction.

Cages placed on old car tyres and stones that are difficult to clean.

Drains and facilities in need of repair.

Wherever possible, surfaces in the LBM (i.e. floors, walls, cages, cage supports, ceilings) should be made of materials that can be easily cleaned and disinfected. If not, a pro-gramme of improvement/replacement and/or repair should be introduced.

The only solution for markets such as the one pictured above is to repair and replace broken drain covers and walkways so that they ca n be cleaned properly.

All surfaces washed with water and detergent. Drains should be flushed regularly.

The following picture is an example of an upgraded market with improved surfaces.

In this example, cages constructed of materials that are easy to clean (and cleaned regularly) are placed over a drain that reduces the likelihood of faecal contamination of footwear. Notice the non-porous floor. Trays should either be placed under poultry cages, or poultry cages should be located above drains. Alternatively, cages can be located so that poultry faeces drop onto areas where the public does not enter and that can be easily cleaned.

Poultry reared in metal cages placed over drains.

❝ POULTRY ARE SOLD ON THE FLOOR ❞

Poultry kept on the ground or on cages on the ground

> Poultry in these two examples are either kept directly on the ground or in cages on the ground. Instead, they should be placed on raised slatted matting floors to allow for faeces to fall through, in cages with faecal collection trays between cages, or in cages placed directly over drains.

Poultry should never be kept on the floor of the market. They should be placed on elevated slatted surfaces or in cages so that they are not lying in faecal material. The slats and cages can be cleaned after trading each day and areas where faecal material can accumulate under cages and slats need to be cleaned and disinfected regularly (see picture below).

Poultry on elevated slotted flooring that reduces faecal contamination.

❝ I HAVE CAGES IN MY MARKET, BUT FAECAL MATERIAL IS NOT COLLECTED ❞

If cages are installed in markets, metal or plastic trays should also be installed below them to collect faecal material and to prevent faeces dropping onto birds in the cages below. If cages are only single level then use metal trays (see photo) or allow faecal material to drop into drains below the cages.

Where possible, measures need to be taken to prevent faecal contamination of areas where customers walk. Note the trays under cages (see arrow) that prevent faeces from falling onto the ground. Note: The use of lime, as applied here, is not recommended; (unless activated by wetting it is not an effective disinfectant and poses an unacceptable risk of combustion).

Systems that prevent contamination of common areas of markets with poultry faeces.

❝ SOME POULTRY IN MY MARKET ARE KEPT ON LITTER ❞

The practice of keeping birds on litter in markets should be phased out as it can result in birds being kept for an extended period of time. If litter is used it should only be used for a single batch of poultry and then discarded. Note: Wooden fences, as pictured here, are very difficult to clean and disinfect. Once litter has been removed, the surface under the litter must be cleaned thoroughly.

Poultry on litter in a market.

❝ DRAINAGE IN MY MARKET IS POOR AND EXISTING DRAINS ARE OFTEN BLOCKED ❞

Facilities for solid waste and liquid runoff, such as drains with solid waste traps should be in place (see picture). These allow waste to be handled in a way that does not pose a risk to the public or to other poultry (see section on cleaning and disinfection in Annex 1). All solid waste should be treated before leaving the market (e.g. composting) or taken directly in closed containers to nearby sites with suitable facilities for appropriate treatment.

Drains that are easily blocked with organic material.

Drains should contain solid waste collectors that are emptied regularly.

Handwashing facilities with soap and appropriate signage should be available at convenient locations in all LBM markets.

❝ I DON'T HAVE ANY FACILITIES FOR HANDWASHING ❞

Clean water and soap should be provided for traders and consumers to wash their hands after contact with poultry. These facilities should be accompanied by appropriate signage, as in the photo below, and appropriate training.

" MY MARKET WATER SUPPLY IS POOR AND I HAVE NEVER HAD THE WATER QUALITY TESTED "

Provide a readily accessible supply of potable water that can be accessed by stallholders and those engaged in cleaning the market and associated equipment. If water supplied to the market has never been tested, arrange for water quality testing. Poor quality water can interfere with cleaning and disinfection.

" I HAVE NO RECORDS THAT ALLOW ME TO TRACE POULTRY INTO AND OUT OF THE MARKET "

Each trader should keep records of the source of all poultry; (this can be difficult if poultry move from market to market – a practice that should be discouraged).

LES SIMS

Tracing system in a retail market.

The picture demonstrates one simple method of tracing that allows birds in markets to be traced back to the farm of origin.

Some form of traceability is essential. In this case, the relevant part of the health certificate for the poultry is placed in a plastic bag on top of the cage housing the poultry, allowing the source of the birds to be traced. Market managers should also keep records of all inward and outward consignments.

" MOST OF MY MARKET STALLS HAVE LOTS OF EXTRANEOUS EQUIPMENT AND MATERIAL IN THEM "

LES SIMS

Excess material stored in market stalls making it difficult to decontaminate.

Stalls should be kept free of unnecessary equipment with separation of keeping/display and slaughtering areas.

Where possible, remove unnecessary items from stalls on a permanent basis to facilitate cleaning and disinfection. Replace temporary canvas roofing with a metal roof. Over time, replace bamboo cages with plastic cages that can be easily cleaned.

❝TRADERS IN MY MARKET SELL DIFFERENT SPECIES OF POULTRY TOGETHER IN THE SAME MARKET STALLS, AND EVEN IN THE SAME CAGES ❞

Examples of mixed species trading in retail markets.

Species should be segregated, whenever possible. As a minimum requirement, different species of poultry should be housed in different cages and preferably in separate stalls. This applies, in particular, to aquatic and terrestrial poultry.

Ideally, different species (especially aquatic and terrestrial poultry) should be segregated and not sold in the same market. In many countries, this segregation is not practised and, until separate markets are established, different species of poultry should be kept in separate cages and, preferably, in separate parts of the market, so as not to allow direct contact between birds of different species.

❝ IN MY MARKET, TRADERS KEEP POULTRY IN THEIR STALLS FOR A NUMBER OF DAYS AND WE DO NOT LIMIT THE LENGTH OF STAY IN THE MARKET ❞

Retention times of poultry in markets should be kept to a minimum. Ideally, no birds should remain in a market for longer than 48 hours (the shorter the retention time, the better). This includes birds that are brought to a market, taken to another place (such as a traders' yard) and returned to the original or another market – practices that should be discouraged.

❝ SLAUGHTERING AND BUTCHERING ARE DONE ON THE FLOOR OF THE MARKET ❞

This practice should be phased out over time after determining the reasons why it is used. It is important to understand the needs of those undertaking this work, as even if more appropriate facilities are provided, old practices will prevail.

Dressing of poultry carcasses is often undertaken on the floor in retail markets.

〝 I DON'T HAVE ANY RULES ON DISPOSAL OR BUTCHERING OF DEAD POULTRY 〞

A dead bird in a live poultry market (on top of the cage) not handled correctly, allowing contact between dead and live poultry.

Sick birds that have been brought to a market as part of a batch of poultry. The birds have been isolated temporarily, but proper facilities for isolation and destructionof sick birds are required.

Bins for disposal of solid waste and dead birds in a retail live poultry market.

Dead chickens must be thrown away into appropriate bins, not butchered or sold. They should be removed from places where others can come into contact with them, to avoid transmission of pathogens.

Traders should not sell or butcher sick poultry or birds that have died.

Any unusual mortality in poultry in markets or transport vehicles must be reported to market authorities or local veterinary authorities.

❝ I ALLOW OTHER ANIMALS TO BE SOLD WITH POULTRY IN THE MARKET INCLUDING RABBITS, DOGS AND YOUNG CHICKS AND DUCKLINGS ❞

Young ducklings for sale in poultry markets selling mature meat birds.

Rabbits for sale in a retail poultry market.

Sale of meat birds and young birds for production in the same market space should be avoided to prevent transmission of disease from mature birds to younger birds and back to farms.

Only poultry should be offered for sale in poultry areas – in the photo above, rabbits are also being offered for sale. The potential for infection of mammals in a market should be minimized.

❝ CUSTOMERS ARE ALLOWED TO TAKE LIVE POULTRY OUT OF THE MARKET ❞

LES SIMS

For retail markets, customers should be discouraged from taking live poultry home by promoting hygienic slaughter; (this will not be feasible in very small local markets with no slaughter facilities). If sale of live birds occurs it is essential to have public awareness and training programmes on ways to minimize risk of disease transmission as a result of this practice.

Live poultry being removed from a retail market.

❝ I DON'T HAVE ANY MARKET EDUCATION PROGRAMMES ❞

Education programmes should be conducted regularly to advise traders on the risks associated with trade in live poultry and ways to minimize these risks. Ideally, these programmes should be designed in conjunction with traders to ensure the content is appropriate to the target audience.

❝ I DON'T HAVE ANY MEASURES IN PLACE TO CONTROL TRADERS WHO BRING POULTRY TO THE MARKET ❞

Currently, traders move from village to village to buy poultry. They then send consignments of mixed poultry either to their own premises, to other trader's yards, or to wholesale or retail markets.

This system can result in a high risk of introduction of zoonotic avian influenza viruses to traders' yards and markets.

A system of regulation of traders should be implemented and, where possible, traders' holding areas should be required to have appropriate facilities for keeping birds. Preferably, there should be a system of segregating birds kept on the premises from any birds returning from markets. Like markets, these premises should be subject to regular cleaning. There should also be poultry-free days during which facilities are cleaned so that the cycle of infection in poultry housed in the holding areas is broken. These requirements may be difficult to implement and enforce in some places.

❝ I DON'T HAVE ANY CONTROLS ON THE MANNER OF TRANSPORTATION OF DRESSED CARCASSES LEAVING THE MARKET ❞

Some existing systems of transport of dressed carcasses from slaughter areas to points of sale result in considerable cross-contamination of carcasses. Measures should be introduced to minimize cross-contamination either by using plastic bags for dressed carcasses or by other means of separation (see Annex 1).

Dressed poultry carcasses being transported from a market to retail outlets resulting in cross-contamination.

❝ I DON'T HAVE ANY RESTRICTIONS ON KEEPING POULTRY OVERNIGHT IN THE MARKET ❞

At the end of the day, any unsold birds should be kept in cages in the market, or in a separate, clean area away from other poultry. If they are taken back to a trader's or farmer's house (practices that should be discouraged) they should be kept separately and not mixed with poultry in these households. Storage areas in these locations must also be cleaned and disinfected and there should be regular rest days in which no poultry are kept.

❝ I DON'T IMPLEMENT POULTRY-FREE REST DAYS IN MY MARKET ❞

Wherever possible, regular and fixed poultry-free rest days should be implemented (minimum of once per month) during which the poultry section of the market is thoroughly cleaned and disinfected and left vacant for 18 to 24 hours (except in markets where overnight keeping does not occur). Communication messages to the public must be in place about the reasons for rest days.

The two main objectives of poultry-free rest days are to allow thorough cleaning of the market without the presence of poultry (which facilitates cleaning) AND to break cycles of infection in poultry housed in the market(s).

Both elements should be included, otherwise the rest day will not minimize the risk of transmission and persistence of zoonotic avian influenza viruses, especially in markets that, otherwise, keep poultry continuously.

Rationale for poultry-free rest days

LBMs receive poultry from different sources. Some markets are managed in such a way that poultry remain in the market for periods longer than the incubation period of the influenza virus, so that the virus can perpetuate.

Cleaning and disinfection help to remove environmental contamination, but if infected poultry are not removed from the market, the environment will very quickly become re-contaminated.

Market rest days for H5N1 highly pathogenic avian influenza were first introduced in 2001 in Hong Kong SAR following detection of the virus in markets and outbreaks of disease in poultry in the markets.

Poultry-free rest days reduce, but do not eliminate, risk of infection with these viruses. There is always a trade-off between the needs of poultry traders (impact on trade of rest days if too frequent) and the effects of the rest day (reduced risk of transmission of disease). As has been proposed for 'platinum' standard markets, the best method of management to prevent markets from remaining infected is to allow no overnight keeping in markets by slaughtering/selling all remaining birds by a set time. However, this can create challenges for traders if demand for poultry is difficult to estimate. If required to slaughter the poultry that remain at the end of the day, traders can suffer losses because of the price differential between a live and a slaughtered bird, paid by consumers.

At this stage, there is limited information on the effectiveness of well-managed rest days on prevention of infection with influenza A(H7N9) virus. Based on experiences with other avian influenza viruses, at least one poultry-free rest day should be implemented every month, but more may be required.

The need for additional rest days can be assessed by testing markets on a regular basis just before, and at various times after, rest days. Such assessment can determine whether markets were contaminated with zoonotic avian influenza viruses prior to the rest day and how quickly markets become re-contaminated. If human cases of infection with H7N9 subtype virus linked to a particular market occur in the period between rest days this is also a strong signal to suggest that either the rest days are not being implemented properly, or that the time between rest days is too long.

Poultry-free rest days should ideally be coordinated across whole cities, provinces or regions to avoid movement of poultry from one market to another, to allow residents to get used to the idea and to allow for enforcement of illegal trading. On these days, it should still be possible to purchase chilled poultry carcasses from central slaughtering facilities.

After the poultry-free rest day, no poultry that were kept in markets previously should return to the markets. Allowing return of birds that have been in markets negates the effect of breaking cycles of infection in poultry.

Ideally, any poultry coming to the market on reopening will have tested negative for evidence of infection with zoonotic avian influenza viruses, but in most countries this testing is not feasible.

If wholesale and retail markets operate in the same areas then it is preferable for wholesalers to reduce the number of birds they order the day before the rest day to ensure that they are not left with poultry in their stall when the rest day commences. Wholesale traders need to plan for and anticipate reduced demand by retail traders in the few days leading up to the rest day. Retailers should plan for rest days by ensuring they have few birds on hand prior to the rest day. The timing of rest days should be determined after discussions with traders, so that the effects on sales are minimized.

The rest day can be staggered slightly so that wholesale markets close and reopen 12 to 18 hours before retail market stalls to ensure there is a supply of new poultry for retail stalls when they reopen.

When done properly, a cut-off time needs to be established for retail markets after which trading will cease and any live poultry remaining in markets must be slaughtered and processed.

For wholesalers, arrangements should be made with slaughterhouses to process any remaining poultry that are in excess of orders from retail stalls.

Overcoming resistance to poultry market rest days

Resistance to poultry-free rest days is known to be high in some countries and reflects a lack of understanding by traders and, in some cases, by government officials of the risk posed by markets that never close. It also reflects the fact that traders are satisfied with existing hygiene standards; ("I have been in this business for 20 years and never got sick"). Time needs to be taken to work with traders so that they understand the reasons for rest days.

Even if market rest days are proposed and implemented, many traders would simply choose to take birds out of the market to other sites, returning with the same birds after the rest day.

Once agreement is reached on implementing rest days, market management personnel must be given sufficient powers to enforce rules related to market rest days, and should be backed by local veterinary and public health authorities. Neither of these is guaranteed at present.

Often, it is not until there has been a crisis in confidence in poultry products that traders accept market rest days as good standard practice.

Although poultry-free rest days are of more benefit for markets in which virus can persist because of management practices, rest days should be implemented for all types of live poultry markets regardless of size, to allow for thorough cleaning and possible disinfection. The exception is markets that do not allow overnight keeping because they are cleaned on a daily basis. (Although in Hong Kong SAR, where there is no overnight keeping, these markets are also subject to a rest day).

During market rest days, thorough cleaning of markets is possible. The focus should be on items that come in direct contact with poultry. However ceilings, walls, floors and equipment should all be cleaned thoroughly (see Annex 1). Standard operating procedures should be developed and those conducting the cleaning, should be trained in how to clean and disinfect items effectively without endangering public health.

Communication on the dates of rest days to the public should be provided well in advance.

❝ I DON'T HAVE A REGULAR CLEANING AND DISINFECTION PROGRAMME AND/OR I DON'T KNOW THE CORRECT WAYS TO CLEAN AND DISINFECTANT MY MARKET, OR THE PROPER WAYS TO USE DISINFECTANTS ❞

Avoid inappropriate use of disinfectants through practical training and development of appropriate operating procedures. Poultry should not be sprayed with disinfectant as it is both ineffective and may adversely affect the health of the birds.

Regular cleaning (minimum daily) and thorough cleaning with additional disinfection (minimum weekly) should be implemented, preferably when the rest of the market is no longer open to the public. This cleaning requires a supply of potable water, cleaning equipment and appropriate personal protective equipment (PPE) for cleaners.

A full guide to cleaning and disinfection is provided in Annex 1.

FAO/ASTRID TRIPODI

Spraying of poultry with disinfectant.

Resources

REFERENCES

FAO/ECTAD. 2010. *Guide for good biosecurity practices in live bird markets (in French).* Bamako, FAO. (available at http://www.fao-ectad-bamako.org/fr/Guide-des-bonnes-pratiques).

FAO/OIE/World Bank. 2008. *Biosecurity for highly pathogenic avian influenza: Issues and options.* Rome, FAO. (available at http://www.oie.int/doc/ged/D5896.PDF).

WHO. 2006. *A guide to healthy food markets.* Geneva, WHO. (available at http://www.who.int/foodsafety/publications/capacity/healthymarket_guide.pdf).

 железо

Annex 1
How to decontaminate live bird markets

BASIC PRINCIPLES

Decontamination is a crucial part of market biosecurity and should be undertaken on a regular basis to remove and inactivate zoonotic avian influenza viruses and other pathogens that might be in the market environment. The major step is initial cleaning that, for influenza viruses, also results in decontamination/disinfection to the point where it may not be necessary to use other disinfectants. It is the most important step in the decontamination process.

Cleaning will remove most disease-causing agents through removal of contaminated organic material. When detergents are used, the cleaning process will inactivate avian influenza viruses.

Application of additional disinfectants can be used as a final step to ensure destruction of any residual virus after cleaning. Many guidelines for market hygiene and disease control recommend use of disinfectant although the effect of detergent on influenza viruses should not be underestimated. In many cases, the use of disinfectants is unnecessary if the goal is elimination of avian influenza viruses and cleaning is done properly. Use of disinfectant is only effective if applied after thorough cleaning as most disinfectants do not work well in the presence of organic matter.

Some markets are difficult to clean because surfaces are not sealed (e.g. earthen floors, broken concrete) or the materials used for holding poultry are not easy to clean (such as bamboo or wire cages). As a rule, the solution to this problem is to improve the quality of surfaces and equipment because any attempt at decontamination of these sites will be hindered by the quality of the facilities. Ensuring that there are no poultry in the market during the process facilitates decontamination.

Cleaning requires an appropriate supply of clean potable water. Water quality should be assessed on a regular basis for total bacterial counts and organic matter. If either of these is too high, appropriate measures need to be taken such as chlorination or filtration or use of alternative supplies to ensure suitable water is available.

Water used for cleaning of poultry areas should not be discharged untreated because it could spread viruses to other virus-free or naive locations. However, this is the situation in many existing live poultry markets. Ideally, holding tanks that allow water to be treated before release should be installed (see section on liquid waste).

Systems need to be in place for management of solid waste picked up during the cleaning process so that it does not pose a risk of transmission of virus (see section on solid waste).

LES SIMS

Disinfectant pit at the exit from a live poultry market.

Cleaning must be conducted in a manner that is both effective and does not adversely affect public health. It requires:

- training of personnel involved in cleaning and, if practised, disinfection. This includes training market traders and market staff in proper use of equipment, detergents and disinfectants, covering the dangers associated with and appropriate handling of disinfectants (i.e. especially concentrates prior to their dilution);
- training in operation of equipment used for cleaning and in cleaning techniques;
- training in fitting and mandatory use of personal protective equipment, especially during disinfection;
- undertaking cleaning and disinfection when markets are closed to the public so as to reduce the likelihood of any airborne particles containing virus from causing infection in market traders or customers. Undertaking cleaning when the market is closed to the public is particularly important if high-pressure washers are being used;
- ensuring any foodstuff in areas adjacent to poultry stalls in wet markets are not contaminated during the cleaning process;
- ensuring appropriate footwear is worn and avoiding slips and falls while areas are being cleaned and disinfected.

Cleaning is the most important step in the decontamination process

In many countries there is considerable waste of disinfectants due to improper use. For example, spraying of poultry in transport and market cages is essentially useless and may cause harm to the birds and the consumer.

Spraying of vehicles with disinfectant without preliminary washing is also of very limited value.

Footbaths and vehicle pits containing disinfectant are usually of limited effectiveness because of the limited contact time with the disinfectant unless they are cleaned out regularly, are topped up with disinfectant regularly and effort is made to clean the surfaces being disinfected before they enter the bath or pit. They do have a psychological benefit in that people become aware of the potential hazards of transporting pathogens into and out of markets (or farms). If footbaths are to be used effectively, footwear should be brushed clean before being placed in the bath.[3]

Training of personnel should include the provision of information on appropriate and inappropriate use of disinfectants.

Cleaning is best conducted as a two-stage process
- dry cleaning by scraping, sweeping, scrubbing;
- wet cleaning by washing with detergents, scrubbing brushes and pressure washers.

When cleaning is completed, **additional disinfection** of each cleaned area can then be done if deemed necessary, using specific, appropriate and safe chemical disinfectants.

While the main objective of cleaning and disinfection programmes in markets is to destroy influenza virus, routine cleaning should be regarded as a risk-reduction measure. Intensive cleaning and disinfection to remove all organic matter is time-consuming and can be expensive, depending on the nature of the surfaces, the equipment being cleaned and the disinfectant used.

In addition, as soon as infected poultry re-enter a cleaned market it will be re-contaminated. Therefore, in areas where zoonotic avian influenza viruses are enzootic, cleaning and disinfection often have only a temporary effect in reducing environmental contamination. Nevertheless, there are very sound reasons to reduce the levels of contamination to a minimum at the end of each day using cleaning routines that result in considerable risk reduction. These should be coupled with regular rest days during which the market can be thoroughly cleaned and disinfected. In both cases, one of the major goals is to reduce the likelihood that any poultry entering the market will get infected as a result of being in the market.

The focus of routine cleaning should be those surfaces that are likely to come in contact with poultry and equipment that has the potential to generate airborne virus, such as defeathering machines.

Special attention should be paid to cages and vehicles that leave the market as these can transfer virus to other virus-free sites. Transport drivers also pose a risk if footwear and clothing are contaminated. Measures should be in place to prevent virus being spread via traders and transporters, as described below.

If routine cleaning and disinfection procedures are too time-consuming or costly there is a very high probability that the measures proposed will not be followed by market traders, resulting in shortcuts being taken. A good example is the cleaning of vehicles, especially if this requires cleaning of both the cabin and the external surfaces. Wherever possible, methods should be simplified to ensure compliance without compromising results. For example, the use of rubber mats in the cab of vehicles allows these to be removed and cleaned reducing the need to disinfect other parts of the floor of the cabin covered by the mats.

Someone independent of the process should monitor cleaning and disinfection. Visual inspections after cleaning provide information on the extent to which organic matter has been removed; (a good rule of thumb for thorough cleaning is that no faeces or feathers should be present once it has been conducted). This is a reasonable indicator of success without the need for microbiological testing.

---

[3] http://www2.dupont.com/Virkon_S/en_GB/applications/disinfectant_foot_dip.html.

There are various other ways to measure effectiveness of decontamination by testing. As a rule, if testing is being done in places with limited resources, it is probably better to test for the target organism(s); in this case, influenza viruses. In most cases, it is not necessary to test given the high susceptibility of these viruses to detergents.

Measuring concentrations of other microbes remaining on surfaces after cleaning could also be done, but only if it is considerably cheaper than testing for avian influenza viruses (for which testing systems are already in place) and if it has been established that the indicator organism is a good substitute for avian influenza virus.

If tests based on amplification of viral nucleic acid are used to assess effectiveness of cleaning and disinfection it needs to be recognized that some disinfectants do not denature viral RNA (an exception being chlorine-based disinfectants and some oxidizing agents). Therefore, both intact and non-infectious virus particles may be detected.

DETERGENTS

Dishwashing liquid (e.g. Dawn) is an ideal detergent because it has been designed specifically to cut through lipids and is non-toxic. The quantity to use depends on the quality of water; (a useful rule of thumb is that there should be soap bubbles in the mix). Ideally, detergent should be used in hot water (preferably 50+°C).

Other detergents such as laundry powders can also be used depending on cost and availability.

Some guidance on quantities for liquid detergents is provided in the following table.

Surface Type	Detergent /100ml water
Very dirty and never been cleaned	15 ml
Dirty but surface is smooth, can be wiped clean	10 ml
Surface is smooth and has been cleaned in last week	5 ml

USE OF DISINFECTANTS OTHER THAN DETERGENTS

Effective decontamination requires cleaning to remove organic matter and may be followed by application of a chemical disinfectant to cleaned, dry areas.

The disinfectant chosen must be effective against the agent in question (which, in the case of avian influenza, includes most disinfectants and detergents) and should not cause environmental or other safety concerns.

For example, glutaraldehyde, mixed usually with a quaternary ammonium compound, is a very effective disinfectant that is widely used in markets in Asia, but it has the potential to affect human health due to sensitization that occurs in some people exposed to the chemical. In addition, it is not suitable for use as a surface disinfectant for food contact areas such as those where poultry are dressed after slaughter.

Other than ensuring surfaces are properly cleaned (which often then means the use of disinfectants is not required to inactivate avian influenza viruses), three other factors must be considered to ensure the disinfectant works properly: the dilution rate, the application rate and the contact time.

If any one of these three factors is not followed properly then disinfection will not be effective.

Dilution rate: This is the quantity of disinfectant added to a set volume of water. The dilution rate depends on the purpose and the disinfectant used. Always read the label of any disinfectant to ensure it is applied at the appropriate dilution. Standard operating procedures and training in these procedures should be developed for each disinfectant to ensure personnel involved in their preparation use the appropriate concentrations and know how to mix them safely as some concentrates are potentially toxic. A calculator, as found on all mobile phones, can help you to work out the quantities needed. For example, if the instructions say you need to add 500 ml of concentrated disinfectant to 25 litres but you have a 7.5 litre container, then you would start with 7 litres of water to allow room in the container for the disinfectant. To determine the amount of concentrate needed divide 7 litres by 25 and multiply the result by 500 ml. By this calculation, 140 ml of concentrate would be needed.

Application rate: The application rate is the quantity of disinfectant solution to be applied to a surface. Most disinfectant solutions are applied at 300 ml per square metre but this depends on the nature of the surface being disinfected. For example, if used on cages, disinfectant should be sprayed until all surfaces are wet. Again, a calculator can help you to work out the required quantities by measuring the width and length of the area to be disinfected and multiplying the two for the area.

Contact time: The time the disinfectant stays on the surface is the contact time. Although some disinfectants can inactivate pathogens rapidly, it is best to allow 30 minutes before washing off the disinfectant. In some situations it is possible to leave the disinfectant on the surface and allow it to dry. Whether or not it is necessary to wash off disinfectant depends on the chemical used.

Some disinfectants lose their potency over time once diluted, especially chlorine-based disinfectants. For this reason, fresh working dilutions of disinfectant should be prepared each day from concentrates. Once again, it is important to follow the manufacturer's instructions if using a commercial product.

MATERIALS NEEDED FOR CLEANING AND DISINFECTION

The following materials are required for cleaning and disinfecting markets. They need to be available whenever a market is cleaned and, therefore, must be kept in or near the market.

- detergents for cleaning, including powder detergents that can be used for scrubbing or handwashing of items, and/or liquid detergents, which are used in power sprayers;
- scrubbing brushes;
- an appropriate supply of the disinfectant (if one is used);
- buckets;
- shovel;
- brooms;
- scraper, preferably with a rubber blade;
- PPE – gloves, apron, boots, mask, plastic glasses or face shield;
- bins for temporary holding of solid waste;
- high-pressure washers, where available.

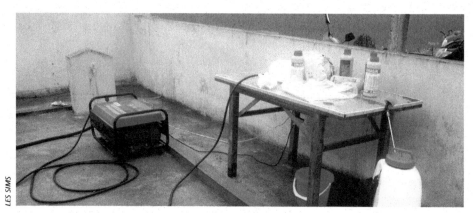

LES SIMS

Equipment used for decontamination including disinfectants and sprayers.

LES SIMS

Scraper, shovel, broom and bin for solid waste storage.

FAO INDONESIA

Equipment used for decontamination.

FAO INDONESIA

PPE, including a face shield, worn by workers using a high-pressure sprayer to clean a market.

Whatever disinfectant or detergent is used, it is important to follow the manufacturer's direction.

This guide does not provide specific recommendation on the chemicals to use as disinfectants. A range of chemicals is available and most commercial disinfectants are effective against avian influenza virus, if used appropriately on cleaned surfaces.

In all cases, disinfectants should be applied to dried surfaces because water present on the surface can dilute the chemical in the disinfectant solution. This mode of application may be difficult to achieve during wet seasons in open markets.

NOTES ON SELECTED CLASSES OF DISINFECTANT

Notes on several classes of disinfectant are provided below.

Additional information on classes of disinfectant and their relative advantages and disadvantages is available in the *Ausvetplan decontamination manual*[4] and in other guides.[5] The following agents are listed because they are generally regarded as the safest to use given health hazards of some (i.e. glutaraldehyde and formaldehyde) or explosion/fire hazards of others (i.e. lime).

Bleach (sodium hypochlorite)

Provided its limitations are recognized, household bleach can be an effective disinfectant for avian influenza virus on cleaned surfaces. It is included here because it is often the most readily available and affordable disinfectant in resource-limited settings.[6]

Cheap and readily available, bleach generally works well to inactivate influenza virus provided working solutions are made up on a daily basis, and if surfaces on which it is used are thoroughly cleaned before application.

However, because it is inactivated readily when in contact with organic matter, some jurisdictions have not recommended its use because of the potential for variation in effects.

Concentrated bleach is corrosive and, therefore, personnel making up working solutions must wear gloves and appropriate eye protection. It is potentially corrosive for metals when used in high concentrations.

It is prepared at the following concentrations depending on the end use:

- 1:10 dilution (1 part bleach into 9 parts water) for areas expected to have had major contamination or for equipment;
- 1:100 dilution (1 part 1:10 dilution into 9 parts water) for general use.

Oxidizing agents (e.g. Virkon-S®)

Oxidizing agents are widely used and are generally non-toxic and effective disinfectants that rely on the oxidizing properties of active ingredients. They are more expensive than household bleach, especially those that are formulated to contain other chemicals.

Oxidizing agents are supplied as a powder that can be applied directly to contaminated surfaces, but they are usually prepared as a solution, in accordance with the manufacturer's guidelines.[7]

Solutions are usually more stable for a longer period than bleach but it is recommended that diluted working solutions should be prepared daily. They are less severely affected by organic matter than bleach.

Acids and alkalis

A number of acids and alkalis can be used as disinfectants and their action depends on obtaining the appropriate pH level. They may be of some value in decontaminating liquid waste.

---

[4] Animal Health Australia. 2008. Ausvetplan operational procedures manual – Decontamination (available at http://www.animalhealthaustralia.com.au/wp-content/uploads/2011/04/DECON3_2-01FINAL12Dec08.pdf)

[5] Disinfectants 101 (available at http://www.cfsph.iastate.edu/Disinfection/Assets/Disinfection101.pdf.

[6] http://www.who.int/csr/resources/publications/surveillance/Annex7.pdf).

[7] See for example http://www2.dupont.com/Virkon_S/en_GB/applications/.

Vinegar is a cheap and readily available disinfectant that could potentially be used to reduce viral contamination on unsealed and sealed surfaces, applied undiluted.

When making up working solutions, it is important to add the chemical to water and not water to the chemical to avoid adverse chemical reactions.

Quaternary ammonium compounds

These are effective against influenza viruses and are widely used as disinfectants. They also have detergent properties.

Ionophores

These chemicals are also capable of destroying influenza viruses. They may require longer contact time and may cause staining.

Annex 2
Practical market decontamination

This section examines each of the sections of live poultry markets and provides guidance on techniques that should be adopted for cleaning for each area. Standard operating procedures should be developed for each market, providing details of all steps undertaken for routine cleaning, additional disinfection and special cleaning on poultry-free rest days.

VEHICLE UNLOADING AREA

Vehicle unloading area in which there is access to the public.

Unloading of poultry in market stalls in a wholesale market.

This area needs to be cleaned after each vehicle is emptied as it will become heavily contaminated with poultry faeces when cages are removed.

In all poultry markets, poultry are unloaded from vehicles either in cages, individually, or in groups of birds. The unloading area usually becomes contaminated with poultry manure and should be cleaned on a regular basis. In some markets, vehicles drive to stalls and unload directly into cages in the stall.

Where possible, solid waste from this area should be removed after each vehicle has been unloaded. The solid or semisolid waste can be stored in bins until it can be treated (see section A2.10).

After all vehicles have unloaded, the vehicle unloading area should be cleaned by first removing solid waste. At the very least, this area should be cleaned once every 24 hours.

The vehicle unloading area should be constructed of material that can be scrubbed (such as concrete) and should have an appropriate drainage system.

All vehicles, cages and other equipment used in this area must be removed from the unloading area during cleaning and should not be returned until they have been cleaned.

Once solid waste is removed, a hose or buckets of water is/are used to wet the area followed by application of some detergent (laundry detergent will suffice or dishwashing liquid can be added to water used for washing). The area should be scrubbed clean. If a pressure washer is available, wash the area with settings of 2 percent detergent rate and high pressure on the nozzle. After cleaning, the detergent should be removed by washing with clean water.

If disinfectant is used, this is applied following the washing and scrubbing. It should only be applied to cleaned areas using standard rates of application of the diluted disinfectant for a minimum of 30 minutes. Depending on the disinfectant used it can be allowed to dry or be washed off.

CRATE CLEANING AREA

Various options are available for transport crates, including cleaning using a pressure washer. Cage washing machines are also available and may be suitable for places with sufficient capital to invest. Other options include the use of dunk tanks: one for initial cleaning and one for disinfection once all of the organic material has been removed.

The crate cleaning area should be located away from public areas and should have a concrete or other non-porous surface. The surface should be cleaned regularly in a manner similar to the vehicle offloading area.

It may not be possible to wash the crates within the market. An alternative, nearby site may be required for both crate and vehicle washing.

Crate cleaning using high-pressure washer (i.e. Karcher or similar).

- After the birds have been unloaded, the crates should be moved to the washing area.
- Add detergent to the pressure washer following the instructions on the label.
- Set up the pressure washer according to manufacturer's instructions.
- Various options are available for cleaning but it is essential to clean all surfaces, both inside and out. Cleaning cannot be achieved fully if crates are stacked before washing the top and bottom of the cages. Once the top and bottom surfaces are

cleaned crates can be stacked so that the other surfaces (inside the cage and sides) can be cleaned.

- All surfaces should be soaked with detergent using the "chem" setting on the pressure washer.
- Allow contact with the detergent on crate surfaces for at least one minute;
- Then, clean all surfaces using the high-pressure fan jet setting on the pressure washer until all grime has been removed.
- The high-pressure pencil jet setting should be used to remove resistant material.
- If all of the grime is not removed using the pressure washer, some surfaces may need to be manually cleaned with a hand brush and detergent.

Disinfection phase
- Use a hand sprayer or mechanical sprayer to apply disinfectant solution.
- Spray from the top crates down to the bottom crate.
- Each standard plastic crate will require about 500 ml of disinfectant solution with sufficient disinfectant used to wet all surfaces.
- Disinfectant can be allowed to dry on the cages.

Dunk tanks
Dunk tanks can also be used for cleaning and/or disinfection of small quantities of cages.

After tipping out all solid waste material each cage should be hosed, preferably

with a pressure sprayer, then dunked in a tank containing detergent solution for a minimum of 30 seconds, removed and scrubbed until all surfaces are cleaned. The cage should then be rinsed with clean water before being immersed in a tank containing disinfectant solution. Care needs to be taken not to introduce organic material into the disinfectant tank as this will reduce the efficacy of the disinfectant. Detergent and disinfectant will need to be recharged regularly with the frequency depending on the degree of buildup of organic matter in the tank.

Dunk tanks for cleaning and disinfecting transport cages.

Automated cage cleaners
A number of commercial companies manufacture automated cage-cleaning equipment. These require considerable capital for purchase. If cages are heavily soiled, a preliminary washing phase with a high-pressure sprayer may be of benefit before placing cages in the washing machine.

Cage washers can include both cleaning and disinfection steps.

VEHICLE CLEANING

The vehicle cleaning area may be located in the market or at some other, nearby location. It should have a sealed surface with appropriate drainage.

The vehicle tray and walls

- All items should be removed from the vehicle tray.
- Solid waste should be removed from the flatbed with shovel, broom or rubber scraper and placed in a solid waste bin.
- Add detergent to pressure washer following the instructions on the label.
- Soak all surfaces of the vehicle with detergent using the "chem" setting on the pressure washer; the focus should be on the areas contaminated by poultry.
- Allow contact with detergent on surfaces for at least one minute;
- Vehicle surfaces, in addition to the chassis, wheel wells and wheels of the vehicle should be cleaned using the high-pressure fan jet.
- Remove excess liquid from the truck flatbed using a rubber scraper.
- Apply appropriate disinfectant to the tray and under the chassis of the vehicle.

The cabin

- Rubber mats should be removed from the cabin, washed with detergent and disinfected.
- Pedals should be cleaned with a wet cloth and subsequently wiped with disinfectant solution.
- Apply disinfectant to all non-metal surfaces within the vehicle cabin, e.g. 1 percent Virkon-S® disinfectant solution or equivalent using a cloth moistened with disinfectant.
- Return cleaned and disinfected floor mats.
- Wipe the steering wheel with a wet cloth and disinfectant solution.

The driver

Drivers should clean and disinfect their footwear with detergent and a scrubbing brush before re-entering the cab. They should always wash their hands before entering their vehicle.

Drivers who have been handling poultry should have a change of clothes that they put on when leaving the market.

LIVE BIRD HOLDING AREA

The live bird holding area is a very important area for cleaning in markets. It is an area where contamination is likely to be high and where virus can persist in poultry, especially if poultry are kept in this area continuously and for periods beyond the generation time of the virus.

Ideally, facilities in the holding area should be constructed of materials that can be easily cleaned such as stainless steel/metallic/plastic cages with metal/plastic trays below each row of cages to collect faeces. Alternatively, if a single layer of cages is used then the cages should be located over drains in a manner that does not allow faecal contamination of

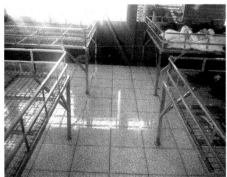
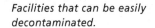

Facilities that can be easily decontaminated.

A wholesale market that is difficult to clean because of the arrangement of cages.

walkways in the market. Birds should not be kept on the floor. If cages are not used, birds should, at the very least, be kept on raised, slatted platforms that allow faecal material to drop to the ground below. These platforms can then be lifted and cleaned once trading is completed.

Some markets use only wire, bamboo or wooden cages on earthen floors and these can be very difficult to clean. This practice should be phased out because it is extremely difficult to decontaminate earthen areas where birds are kept.

Some markets also keep birds on litter, a practice that should be discouraged unless the batch can be sold rapidly. Batches should not be replenished with new birds from different sources and the litter should be removed after each batch.

Litter can be used under elevated slatted floors to trap liquid and solid waste and to reduce the likelihood of liquid run off. The material can then be removed on a regular basis for composting. By using litter in this manner it may be possible to reduce the frequency of cleaning under slatted floors.

Cleaning of the live bird holding area is made difficult if birds are kept overnight in cages. It is best achieved when the cages are empty, which is one of the advantages of poultry rest days and systems of management that do not allow overnight keeping.

Access to the area under cages is required to allow for regular cleaning.

If equipment such as water troughs are attached to cages these should be removed on a daily basis, cleaned and disinfected. Water in water troughs should be replaced on a regular basis (several times per day) as contaminated drinking water can be a source of infection.

Cages should be washed regularly and, preferably, every day at the end of trading, but in many markets this is not possible because poultry are kept overnight. Trays under cages should be scraped clean daily and washed regularly.

Any solid waste in the holding area should be collected and placed in bins for subsequent disposal, including any litter on which poultry have been kept.

Once solid waste is removed:

- Clean cages and any slatted plastic flooring with water and detergent, and scrub with a brush if required until there are no residual feathers or faeces. Allow to dry and, if required, spray the items with disinfectant so that all surfaces are wet.

- On rest days, or when there is evidence of contamination with faecal matter, walls should be cleaned by spraying with water and detergent, and scrubbing should be done as required to remove any buildup of faeces or other organic material.
- Floors should be cleaned daily using methods described for poultry unloading areas.
- Apply disinfectant to·the cleaned, dry surfaces using a sprayer at the appropriate dilution.
- Drains should also be cleaned with any solid waste in traps in drains removed on a daily basis, or more frequently if large amounts of solid waste enter the drainage system (see below).

POULTRY PROCESSING AREAS

The poultry processing area can be heavily contaminated with materials from the dressing process, including blood, offal, feathers, fat and faeces. It requires regular and thorough cleaning and disinfection not only to control zoonotic avian influenza viruses, but to ensure carcasses produced in the market are not contaminated during the slaughtering and butchering processes.

The walls and work areas in the poultry stalls and evisceration area should be constructed of materials that are easily cleaned, such as tiles or concrete, and they should be kept in good repair.

- Sweep the walls and collect feathers, dirt and poultry offal into the rubbish bin.
- Move equipment from the area to be cleaned.

Bleeding drum/cone

LES SIMS

Well cleaned bleeding cones in a small scale slaughterhouse.

The bleeding drum/cone is a high-risk piece of equipment, especially for H5N1 virus, as it will contain feathers, blood, upper respiratory tract excretions and, potentially, faeces from slaughtered poultry:

- The drum should be soaked with at least a 4 percent detergent solution. It should be scrubbed with a coarse brush to remove grime from inside and outside the drum. A pressure washer can be used, if available, but it is preferable to use low-pressure cleaning first to avoid producing airborne particles that may contain virus.
- The equipment should be cleaned at the end of each trading day and disinfected when dry using a disinfectant approved for use in food-processing areas.

The scalding tank

LES SIMS

Pans used for scalding (left hand side). Note tiled surfaces and persistence with processing of poultry on the floor.

Pans used for scalding (left hand side). Note tiled surfaces and persistence with processing of poultry on the floor.

- As many carcasses are dipped in the scalding tank during each slaughtering session, it is important that it is emptied, rinsed with clean water and refilled with clean water at the completion of each session.
- Similar to bleeding cones, the scalding tank should be soaked with at least 4 percent detergent solution.
- It should be scrubbed with a coarse brush to remove grime from inside and outside the tank.
- To clean, use a pressure washer, if available, preferably at low pressure;
- Clean equipment at the end of the trading day and disinfect when dry.

ROTARY PLUCKING MACHINE

LES SIMS

LES SIMS

Defeathering machine in a slaughter area well separated from other equipment and with surfaces that are easy to clean.

Slaughter area and defeathering machine that are difficult to clean because of extraneous material in the stall.

- To reduce the buildup of feathers during the day, fresh clean water should be sprayed in the plucker bowl during operation and after each bird is plucked.
- Turn off and disconnect the power supply to the plucker before cleaning.
- After the session is finished, clean the plu cker bowl with detergent, scrubbing brush and fresh water.
- Make sure that the plucking machine fingers are not cracked or broken and replace them, if necessary.
- Wash and scrub the outside of the bowl with water and detergent.
- Note: Hot water mixed with detergent will help remove fat that builds up in the bowl.
- Rinse with fresh water after cleaning.
- Pressure wash the plucker, if a pressure washer is available, but ensure the electric motor is not in contact with water.
- Disinfect the cleaned plucking machine using an appropriate disinfectant. Allow to dry and after 30 minutes rinse with clean water.

Areas used for additional dressing of poultry (clean areas)

Dressing of poultry should be done on a surface that is easy to clean (preferably a stainless steel table) and that is elevated above the floor so that human or vehicle traffic cannot contaminate carcasses. The dressing process should occur on a clean and disinfected surface in an area separate from live bird holding areas and the slaughter area. In many markets in Asia, processing is conducted on the floor. These practices should be changed over time to improve overall food hygiene. There are several necessary steps for cleaning these areas:

Carcass dressing on the floor; this practice should be phased out.

A carcass handling area with water troughs and metal table.

- Remove all solid waste and dispose of in bins kept for this purpose.
- All surfaces in the dressing area should be washed with detergent and scrubbed with a brush, including knives and cutting boards;
- The area should then be disinfected using a disinfectant suitable for food-processing areas, should be left until dry and then rinsed with water before being used again.

WALKWAYS

- Remove any solid waste and sweep clean the walkways, placing all rubbish in a bin for disposal.
- Wash down all walls and walkways with water from a hose or pressure washer.
- Use pressure spray to clean walkways and walls and add detergent to remove grime or built-up organic matter. Leave areas to soak for up to 10 minutes, then rinse off with water from a hose or pressure washer;
- Spray walkways with disinfectant and then wash off with water.

DRAINAGE SYSTEM

A well-constructed and maintained drainage system keeps wastewater separated from poultry, customers and traders. However, in many markets, drains are either non-existent or poorly constructed. They are often incompletely covered, do not have appropriate solid waste traps or contain material that causes blockage such as plastic and organic matter. All markets should aim to have an appropriate drainage system leading to areas where liquid waste can be held before discharge.

LES SIMS

Drains in a slaughter area. These need to be cleaned regularly to remove solid waste.

The drainage system flow should move from the cleanest areas (where the dressed poultry are sold) through the 'dirty' areas (holding, killing and plucking area) to the area outside of the market.

Drains should contain traps for solid waste that are emptied regularly by designated staff.

All drains must be hosed out each day after removal of solid waste in solid waste traps. Care should be taken to avoid blockage of drains with garbage.

As water in drains will eventually need to be discharged, at that point the liquid must meet appropriate discharge standards set by local authorities. Liquid discharge from markets should not pose any hazard to the health of animals outside the market.

SPECIAL CASES

A number of markets are not amenable to cleaning and disinfection because they are located on earthen floors, lack appropriate drainage or the cages used are not easily cleaned because they are made from porous materials.

The emergence of zoonotic avian influenza viruses and greater demands from the public for traceability and improved food safety provide very strong justification for eventual modification of these markets and changes to market chains. However, changes to existing practices will face strong opposition from traders and transporters who have conducted their business in this manner for many years, unless they see the benefits in the changes (such as increased market share because of the measures taken). Public and/ or private investment will be needed. Any changes will also require strong enforcement to limit development of parallel market chains. Such enforcement is challenging in low-re-source settings, especially in places with endemic corruption.

Cleaning and disinfection of porous surfaces
Ideally, all equipment and surfaces in markets should be constructed of non-porous mate-rials that are kept in good repair. Until old equipment and surfaces can be replaced with those made of non-porous materials it is necessary to implement cleaning programmes to reduce the risk posed by these markets.

Cleaning of timber, cracked or worn walls
- Apply pressure washer spray with 4 percent detergent solution and leave for at least ten minutes.
- If a foaming attachment is available for the pressure washer lance, apply foam to all surfaces and leave for ten minutes as the foam will remain on the surface for a long time, loosening the blood, fat and grime from the surface.
- Use pressure washer spray on a high-pressure setting to remove the grime from the surface.
- Apply disinfectant to the cleaned surface at the recommended rate.

Cleaning and disinfection earthen surfaces
Earthen floors are very difficult surfaces to clean and disinfect. In the long term, markets should not be located in such areas.
Options for treatment that might help to reduce viral contamination include:
- removal of solid litter and waste on a daily basis;
- removal of the top layer of soil/mud (usually impractical);
- shifting locations so that the same areas are not used repeatedly;
- use of disinfectants such as weak acids (e.g. vinegar) which might reduce levels of surface contamination, and might be an option for areas that are already wet.

TREATMENT OF SOLID WASTE
Live poultry markets generate large quantities of solid waste comprising poultry manure, straw, litter, feathers and, in some cases, waste from slaughtering.

No solid waste should be removed from a market unless it has been treated appropriately or is being transported directly to a site for treatment in a covered container.

Composting is one of the best methods available for treatment of solid waste and, if conducted properly, will result in inactivation of most pathogens – (exceptions are some spore-forming organisms) – as a result of the heat generated in the composting process.

Composting requires technical expertise and appropriate facilities. Faecal material will require the addition of organic material such as litter or straw to ensure appropriate temperatures are reached to destroy avian influenza viruses and most other pathogens. There are several guides to composting of poultry manure, which are listed in the footnote below.[8] Professional guidance from experts in composting should be sought before installing composting facilities.

TREATMENT OF LIQUID WASTE

LES SIMS

Biogas tank for liquid waste.

Liquid waste should travel along drainage systems to a storage tank that allows appropriate treatment of the waste before discharge. Biogas facilities and septic tanks are possible options, as are appropriately constructed liquid waste treatment systems based on anaerobic and aerobic treatment of waste. Complex treatment systems can be expensive to install and operate and are rarely used in markets. Any waste discharged from markets must meet local environmental standards and should have been treated so that it does not contain any viable influenza virus (or other major pathogens) when discharged. This treatment may require changes in pH of the liquid waste held in holding tanks. Professional guidance from wastewater engineers should be sought when designing liquid waste handling and treatment systems.

---

[8] http://ext100.wsu.edu/king/wp-content/uploads/sites/17/2014/02/Using-Manure-as-Compost1.pdf
http://www.ianrpubs.unl.edu/pages/publicationD.jsp?publicationId=567.
http://www.dpi.nsw.gov.au/__data/assets/pdf_file/0004/140359/Best-practice-guidelines-for-using-poultry-litter-on-pastures.pdf.
http://www.deq.state.ms.us/mdeq.nsf/pdf/SW_PoultryLitterGuide09232009/$File/guide_poultry_litter.pdf?OpenElement.

FAO ANIMAL PRODUCTION AND HEALTH GUIDELINES

1. Collection of entomological baseline data for tsetse
   area-wide integrated pest management programmes, 2009 (E)
2. Preparation of national strategies and action plans for
   animal genetic resources, 2009 (E, F, S, R, C)
3. Breeding strategies for sustainable management of animal genetic resources, 2010
   (E, F, S, R, Ar, C)
4. A value chain approach to animal diseases risk management – Technical foundations
   and practical framework for field application, 2011 (E, C)
5. Guidelines for the preparation of livestock sector reviews, 2011 (E)
6. Developing the institutional framework for the management of
   animal genetic resources, 2011 (E, F, S, R)
7. Surveying and monitoring of animal genetic resources, 2011 (E, F, S)
8. Guide to good dairy farming practice, 2011 (E, F, S, R, Ar, C, Pt^e)
9. Molecular genetic characterization of animal genetic resources, 2011 (E)
10. Designing and implementing livestock value chain studies, 2012 (E)
11. Phenotypic characterization of animal genetic resources, 2012 (E, F^e, C**)
12. Cryoconservation of animal genetic resources, 2012 (E)
13. Handbook on regulatory frameworks for the control and eradication of hpai and other
    transboundary animal diseases – A guide to reviewing and developing the necessary
    policy, institutional and legal frameworks, 2013 (E)
14. *In vivo* conservation of animal genetic resources, 2013 (E)
15. The feed analysis laboratory: establishment and quality control, 2013 (E)
16. Decision tools for family poultry development, 2014 (E)
17. Biosecurity guide for live poultry markets, 2015 (E)

Availability: October 2015

Ar	– Arabic	Multil	– Multilingual
C	– Chinese	*	Out of print
E	– English	**	In preparation
F	– French	e	E-publication
Pt	– Portuguese		
R	– Russian		
S	– Spanish		

The *FAO Animal Production and Health Guidelines* are available through the authorized FAO Sales Agents or directly from Sales and Marketing Group, FAO, Viale delle Terme di Caracalla, 00153 Rome, Italy.

Find more publications at
http://www.fao.org/ag/againfo/resources/en/publications.html